ALLEN PHOTOGRA

GW00738554

FEEDING HORSES

CONTENTS

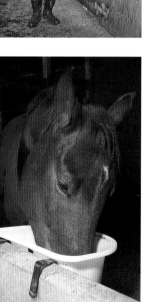

FEEDING STRATEGIES

The more you find out about feeding horses and ponies, the more complicated it can seem. With so many feeds and products available and so much research being done, it may seem difficult to know where to start. The answer is to look at how the horse has evolved and how we can combine that with the demands we put on him.

Horses and ponies are herbivores; they are designed to eat grass and plants and to follow a diet that is high in fibre. In the wild they travel miles each day looking for grazing and water supplies and spend most of their time eating.

By keeping them in small fields where the grass is usually of much higher quality, stabling them and asking them to work, we change their demands for fuel. But to keep them happy and healthy, we must base their management and diet on nature.

This book provides up-to-date guidelines, but each horse is an individual. If you have a problem, consult your vet or an equine nutritionist. The major feed companies have nutritionists who will give free advice, albeit based on their products.

REMEMBER

Forage is a vital part of the horse's diet. Hard feed – mixes, cubes or straight cereals – provide extra energy and can also add fibre to the diet but good quality grass and hay must always be looked on as important building blocks, even though they may need supplementing.

NATURAL BASICS

The horse's digestive system means he needs to 'trickle feed' rather than eat large amounts at infrequent intervals. The breaking down process starts when he chews the food, which then passes into the stomach (1). The stomach is very small and cannot cope with more than about 2.5 kg of food at once. The next digestive stage takes place in the small intestine (2) and the process finishes in the large intestine (3).

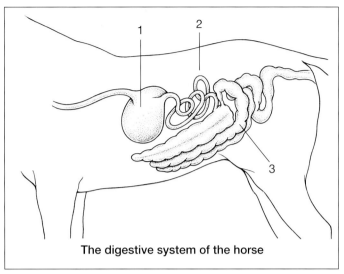

The digestive system of the horse

Routine dental and worming programmes will keep the horse in good condition to digest his food and get the maximum value from it. His teeth must be attended to at least once a year – sometimes more frequently – and your vet will suggest a worming programme. It is also important to look after your grazing.

Ignoring these preventive measures puts the horse's health at risk and is false economy. The risk of colic may be increased and you will have to feed more to try to keep him in good condition.

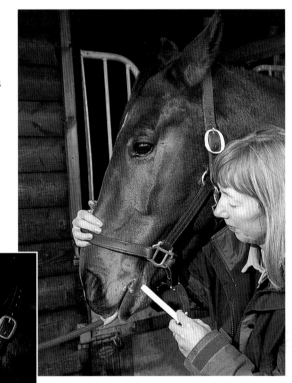

DID YOU KNOW?

Researchers have found that a horse will chew food into pieces as small as 1 mm. Each kilogram of hay will be chewed up to 6,000 times before swallowing.

HOW MUCH FEED?

Many factors affect how much a horse or pony needs to eat, including age, workload, and breed or type. An in-foal mare or one with a foal at foot will have higher energy requirements and a growing youngster may need more fuel than some mature animals. Horses in hard work will obviously have different needs from those at rest or in light work.

Similarly, some horses are better 'doers' than others and seem to live on fresh air – cobs and native ponies are good examples – though they still need essential nutrients. Each horse must be treated as an individual, but the best starting point is to reckon that he will need a daily total amount of 1.5 – 2.5 per cent of his bodyweight.

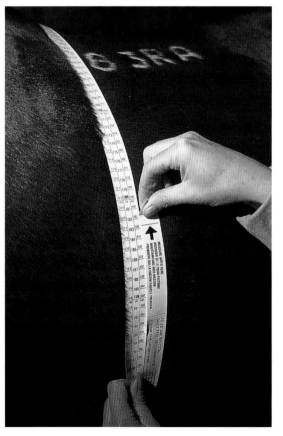

WEIGHTY MATTERS

The most accurate way of finding out his weight is to use a weighbridge, sometimes found at large veterinary practices and racing/competition yards. If this is not feasible, you could use a weightape or mathematical formula. These are not strictly accurate but if used weekly, show whether your horse is maintaining, gaining or losing weight.

FIGURE IT OUT

If you measure your horse's girth and also the length from the point of the shoulder to the point of the buttocks, you can use a mathematical formula to calculate his weight. Measuring in centimetres, the girth squared multiplied by the length and divided by 8717 will give you his rough weight in kilogrammes.

Feed companies often print weight guidelines on the backs of bags, but these have to cover a wide range of animals and can only be general. On average, a 12.2 hh pony weighs about 300 kg, a 14.2 hh 425 kg, a 15.2 hh Thoroughbred about 500 kg and a 16.2 hh hunter about 600 kg.

If the horse is underweight, plan on feeding 2.5 per cent of his bodyweight daily. If he is overweight, limit intake to below 2 per cent but ensure that he still receives essential nutrients. If he is just right, use 2 per cent as your guideline.

The accepted definition of good condition is that the ribs are just covered but can be felt, the rump is rounded and there is no artificial crest caused by fatty deposits. Drastic cases need skilled care: the horse here is pictured before his rehabilitation by the International League for the Protection of Horses and is pictured elsewhere in this book in good condition.

Too much weight puts unwanted strain on heart, lungs and limbs and may predispose the animal to laminitis. Youngsters whose bodies are too heavy for their immature limbs may develop permanent problems.

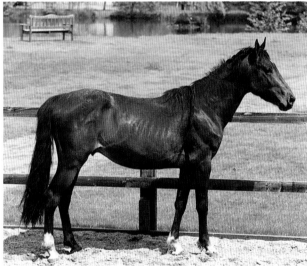

TOO FAT OR TOO THIN?

A horse who is too fat needs more exercise and less food: think in terms of a fibre-based diet supplemented, if necessary, to provide essential vitamins, minerals and so on. Aim for gradual weight gain for an underweight horse and rug him up whenever necessary – otherwise up to 80 per cent of the energy from his feed will go to keeping him warm.

CORRECT RATIOS

You also need to decide the correct ratio of forage to hard feed (concentrates). This depends on the horse's workload, but the forage portion should rarely fall below 50 per cent even if he is in hard work and will often be higher.

Many people overestimate their horse's workload. If he is hacking and doing a little slow canter work he is in light work. Hacking, schooling and competing in dressage and/or show jumping means a medium workload. Hard work really only applies to horses competing regularly in horse trials, cross-country or endurance riding and to some driving animals.

Other factors to take into account include your horse's age and temperament. Growing youngsters, breeding stock and old animals all have different requirements. Feed cannot change temperament, but the right diet can help minimise problems. For instance, a 'fizzy' horse may benefit particu-

larly from a diet high in fibre and sources of slow-releasing energy.

Your feeding programme should be based on his current, not anticipated, workload. Build up his feed, if necessary, as you build up his work – but not in advance.

THE FORAGE FACTOR

OUT TO GRASS

Correctly managed grazing, supplemented when necessary by hay or alternative forage, forms the basis of most equine diets. In average weather conditions, the feed value of grass is highest in spring and early summer and there is usually also a flush of growth in early autumn.

Native ponies and some cob types will need little if any extra feeding when grass is good. Many will need grazing restricted to avoid the risk of laminitis. The latest research has prompted new ideas about the way such animals should be managed, so consult your vet.

The amount of land needed to keep horses depends on its management and whether it is grazed all year round. A rule of thumb is a minimum of one acre per horse. If you divide your paddock into small areas which can be alternately grazed and rested, you will get more out of it.

Pick up droppings as often as possible to reduce the worm burden and get advice from an equine specialist on fertilising. Not all fertilisers are suitable for horse paddocks.

WARNING!

Do not let anyone tip lawn mowings in your field; these ferment and can cause colic. Garden clippings can also be poisonous, as are many plants – including the nightshades, yew, acorns, laurel and ragwort, which is pictured here. Poisonous plants should be pulled up and burned and trees fenced off so they are out of reach.

FEEDING HAY

Horses and ponies need hay when grass is scarce or if they are stabled. The main types of hay are seed hay, made from specially sown seed mixtures, and meadow hay, made from permanent pasture. The quality and nutritional value of any hay is affected by when it is cut and how well it is made and stored.

It should be clean and free from weeds and poisonous plants. Never feed dusty or mouldy hay, as it will affect the horse's health even if it is soaked. Good quality hay is greenish rather than brown, smells sweet and is not dusty when shaken, but the only really accurate way to assess feed values is to have samples analysed.

Many people prefer to soak all hay to reduce the dangers from spores and dust. Nutritionists now say that soaking for ten minutes to an hour, depending on the amount and how tightly it is packed, is sufficient. It is now felt that soaking for 12 to 24 hours, as used to be advised, leeches out the nutrients.

SOAKING SENSE

Always totally submerge each batch of hay in clean water. If you re-use the water you are soaking hay in pollutants. Soaking only helps if the hay is eaten whilst wet: problems return as it dries. Dampening with a hosepipe or 'steaming' hay by pouring on boiling water does not give the same benefits.

Hay can be fed on the ground, in a haynet or from a hayrack. Feeding from the ground is the most natural way, but can be wasteful if the horse treads hay into the muddy ground or his bedding. Haynets avoid waste but must be tied securely: remember that they will hang lower when empty so make sure the horse will not catch his foot in the empty net.

Hayracks, or devices that are a cross between a net and a rack, are a good compromise. If your horse eats as fast as he can, a rack or net with a small mesh will slow him down and he will be occupied for longer. You can buy nets with a small mesh or put one ordinary one inside another to get the same effect.

Whatever method you use, it is best to weigh your hay rather than relying on guesswork – feeding quantities are calculated on dry hay. The easiest way is to hang a haynet from a spring balance. If you use big bales rather than conventional small ones, remember that they are not as tightly packed and a full haynet may not contain as much as you think.

When feeding hay outdoors, put out one more portion than the number of horses. If a horse is chased off by dominant companions he should still be able to find his share.

OLD AND NEW

It used to be thought that hay should only be fed when it was a year old. Nutritionists now say that it can be fed much sooner, as long as it is introduced gradually by mixing a small amount at a time with the previous supply. Hay gradually loses some of its feed value as it gets older.

HAYLAGE

Hay is grass that is cut, dried completely and baled. Haylage – marketed under brand names such as HorseHage – is grass or alfalfa which is baled as soon as it has wilted to about 60 per cent dry matter. It is then compressed and packed in heat-sealed bags, where it ferments in its own 'juices'.

The advantages are that it is almost completely free of spores and dust, so does not need soaking and is ideal for horses with respiratory problems. It has a guaranteed nutritional value, whereas the feed value of ordinary hay gradually decreases. Although it is more expensive, the higher feed value may mean you are able to reduce the quantity of concentrates.

Haylage must be used within three or four days of the bag being opened. If the bags are pierced or chewed by vermin, moulds may develop – if this happens, the haylage must be discarded.

Big bale haylage grown especially for horses is useful for large yards where it can be used quickly enough. Do not feed silage grown for cattle: the way it is made means there is a risk of botulism, which can be fatal.

MEALS ON WHEELS

A small haynet in the horse-box or trailer helps many animals settle on the journey. If you take a spare supply for the journey home, do not tie it to the back of the vehicle or it will be polluted by thrown-up dirt and fumes.

ALTERNATIVES TO HAY

If good hay is scarce and it is not feasible to feed haylage, your horse needs an alternative source of fibre. Clean oat straw, if available, is one option: unlike wheat straw, it is digestible. Wheat straw contains an indigestible woody substance called lignin, and if the horse eats too much it may predispose him to colic.

High fibre alfalfa- or grass-based products are becoming increasingly popular. Many can be fed as a hay substitute, but usually work out more expensive. As hay shortages have become a fact of life in many areas over the past few years, most feed companies have formulated high fibre feeds to try to compensate.

Hydroponic 'grass', grown in trays in indoor units, is another option. Commercial units are available to suit everyone from the one-horse owner to business establishments and though relatively expensive to buy, are claimed to produce a cheaper source of food. They usually produce grass 'mats' from barley seed, which horses seem to enjoy.

HARD FEED

If your horse needs more fuel than forage alone can provide, hard feed – an overall term for cereals, coarse mixes and pellets – can form up to 50 per cent of his total intake. In a minority of cases, such as racehorses and advanced event horses at peak fitness, the proportion of hard feed may be higher but these are the exceptions to the rule.

COMPOUNDS OR STRAIGHTS?

The first decision you must make is whether to feed cereals, traditionally oats or barley and sometimes called 'straights', or commercially prepared compound feeds in the form of coarse mixes or pellets. Most owners prefer compounds, which take much of the guesswork out of feeding; the nutritional content is guaranteed, whereas the value of one batch of cereals may be very different from that of the next.

Coarse mixes (known as sweet feeds in the USA) look more attractive to us than pellets – which may also be called cubes or nuts. However, there is no nutritional benefit although coarse mixes usually take longer to eat and some horses seem to prefer them. Always choose a feed designed for your horse and his actual workload, not what you would like him to do: feeding a competition mix to a horse in light work will not turn him into an event horse!

Most, though not all, compound feeds contain a small amount of molasses to make them palatable and help bind the ingredients. Some people believe that added sugar is as bad for horses as it is for people and opt for feeds with little or no molasses.

If you prefer to feed cereals, your main choices are oats or barley. Maize (or corn) can be fed in small quantities and is particularly popular in the USA, where it is fed on the cob.

Oats are the most traditional feed for horses as they cause few digestive problems and most animals like them. They have always had a reputation for causing ponies in particular to 'hot up' and many people will avoid using them because of this. The latest research shows that a possible reason is that the energy in ordinary oats is released quickly, perhaps coinciding with the times when horses are ridden.

Contrary to popular belief, barley has more energy – weight for weight – than oats. Both cereals have a poor ratio of calcium to phosphorus, which must be compensated for. You can do this by feeding a commercial 'oat balancer', which is equally suitable for use with barley. Alfalfa, or lucerne (shown below), is high in calcium and helps compensate for deficiencies.

DON'T DILUTE

Do not add extra oats or barley to compound feeds. You may think you are giving your horse an energy boost by adding a few oats but the reverse is true, you are actually diluting and unbalancing the feed. Compound feeds are formulated to give the correct balance of nutrients, including vitamins and minerals, when fed at the minimum recommended quantities. Adding cereals destroys the balance.

CEREAL PREPARATION

Oats and barley are traditionally rolled, bruised or put through a cooking process to make them more digestible, though one school of thought now believes that oats should be fed whole. A type marketed as Naked Oats, which has no husks and is higher in oil than ordinary oats, is always fed whole. Energy is released more slowly, so Naked Oats (top) are said to overcome 'fizziness' problems that may be caused by ordinary ones.

Bruised oats or barley are put through rollers so that the husks are cracked. Rolled cereals are put under slightly more pressure and crushed ones are flattened. The risk with crushed cereals is that some of the goodness will be left on the rollers or the floor.

At one time cereals, especially barley, were boiled in an attempt to make them more digestible, but this destroyed some of the vitamins. Modern techniques of extrusion, micronising and steam flaking are much better – they break up the starch molecules, thus improving digestibility, without losing nutrients. Extruded grain is cooked at high pressure whilst micronising results in flakes of cereal. Micronised cereals are often sold simply as 'flaked' barley or maize.

Specialist cooked feeds are sold to promote weight gain. These can be added to any feed, but check the manufacturers' feeding instructions.

ADDED EXTRAS

There are several useful additions that bulk out hard feed – whether it be a compound or straight cereals – without altering the nutritional balance. Mixing it with chaff or chop – terms often used to describe everything from molassed, chopped hay to dried, chopped alfalfa – means it will take him longer to eat, which is usually an advantage for the horse who is stabled part of the time. The longer his food occupies him, the less time he has to become bored!

Carrots or apples are a much appreciated treat. They can also be used to tempt a fussy eater and to give a horse or pony on a restricted intake something to look forward to at feed time. They must be cut up into small enough pieces to prevent the horse choking on them. Carrots should be sliced lengthways. Soaked sugar beet pellets or shreds are high in fibre and a useful source of slow-releasing energy.

Some horses like fodder beet (different from sugar beet and grown mainly for cattle) but others are not so keen. Swedes and turnips also produce mixed reactions – horses, like people, have different likes and dislikes. If your horse enjoys root crops and succulents like these, you can feed up to about 4 kg per day; divide them between each feed, following the guidelines in the Golden Guidelines section.

Soaked sugar beet pellets or shreds add fibre and are a good source of slow-releasing energy. Pellets must be soaked in at least three times their volume of cold water for a minimum of 12 hours; some manufacturers recommend 24 hours. If they are fed dry, they will swell during the digestive process and may cause colic.

Opinions vary on whether sugar beet shreds must be soaked. It is perhaps best to play safe, though they do not need to be soaked for as long as the pellets. Soaked sugar beet quickly goes sour in the heat and will freeze in winter. If you cannot keep it indoors, lining a large container with straw or hay and putting your container of soaked sugar beet inside often provides enough insulation to present freezing. Alternatively, try using an insulated picnic box.

You will probably not need to feed it in summer, but it is a useful succulent in winter.

Bran is a traditional feed that has lost popularity because it is poorly balanced. Bran mashes, made by pouring boiling water on bran to make a sloppy mixture and letting it cool, should not be fed routinely but may be advised by your vet under certain circumstances.

WARNING!

Never risk sugar beet pellets being confused with ordinary feed pellets. Keep containers well apart and label them if necessary.

FEED PELLETS   SUGAR BEET

GOLDEN GUIDELINES

No matter what sort of feed you buy, there are a few golden guidelines to keep in mind. Base your feeding programme around them and you will reap the benefit: a horse who is in good condition for the work he is asked to do. If he is content with his lifestyle, this will also reflect in his attitude to work.

FEEDING SENSE

When working out his diet, take into account his age, bodyweight, temperament and work-load. This will allow you to cater for him as an individual; for instance, one horse may be a good doer whilst another of similar physical type may need more fuel because he is naturally more active.

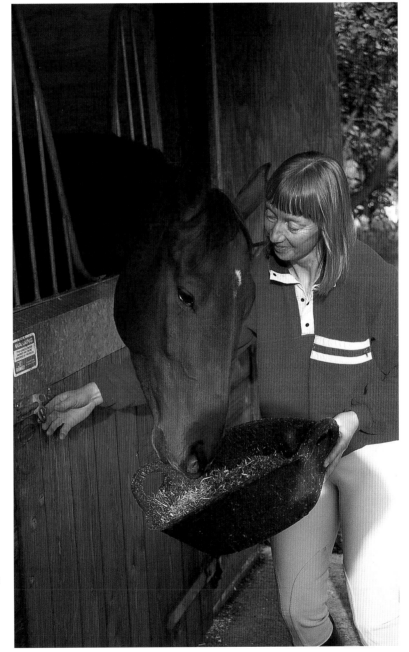

Divide his total ration into frequent small feeds rather than infrequent larger ones. Remember that his digestive system can only cope with a maximum of about 2.5 kg of hard feed at a time: he might munch through twice as much quite happily, but over-loading could cause problems such as colic. Horses like routine, so try to feed at the same times each day and make sure that clean, fresh water is always available indoors and out.

If you introduce a new feed or a new batch of hay, do it gradually over several days. Mix in a little of the new with the old so that he gets used to it. Making a sudden switch could cause digestive upsets and again lead to colic. Each feed should be the same size; the only exception is that some people like to make a stabled horse's last feed of the day slightly larger than the others, as there is a longer gap between it and the next one.

Always feed according to the work he has done, not what he is going to do next day. If his workload decreases, or if you give him a day off, cut down his feed. Ignore traditional advice that working horses need a bran mash once a week – this is the equivalent of giving yourself a weekly laxative whether you need it or not!

Even if you are in a hurry to ride, allow your horse at least an hour to digest his feed before you tack him up. He cannot work on a full stomach any more than you can, and may get colic. Similarly, make sure he has cooled down after work before giving him a feed.

HOT HORSE?

Do not withhold water from a horse before he works; if it is always available, he will drink only when he needs it. If he is very hot, perhaps after a competition, offer half a bucket with the chill taken off immediately and the same twenty minutes later. He should then be able to drink as he wants.

It is always better to feed by weight rather than by volume – or guesswork. A scoop of one feed will not necessarily weigh the same as a scoop of another, so you could be drastically over or under feeding without realising it. Some feed companies make useful measures marked with the weights of particular feeds, or you can make your own.

If your horse leaves any of his feed, throw it away rather than keeping it to offer again; he may turn up his nose even more at 'secondhand' feed and it may go stale. Are you offering too much, or is he off colour? Check his temperature, respiration and pulse rates and take into account his overall attitude. If necessary, get veterinary advice.

Let your horse eat his feed in peace and quiet, even if you are in a hurry. It is not fair to groom him or adjust his rugs whilst he is eating and you may make him rush his food or become bad tempered.

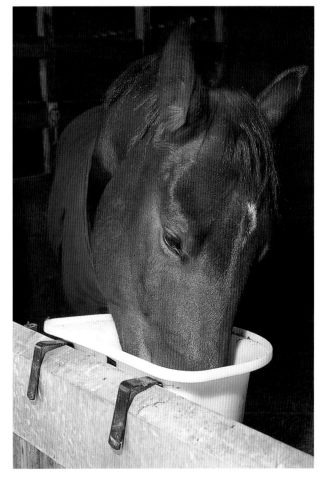

MEAL SERVICE

Some horses prefer to eat at the back of their stables, but those who rush to the door between mouthfuls may be happier with a door manger so they can eat and watch what is going on at the same time. Siting corner mangers in adjoining corners may lead to squabbles between neighbours, so put them in different ones if possible.

WATER MATTERS

Water is just as important to your horse's well-being as feed and makes up 60 per cent of his adult bodyweight. In hot weather, a working animal will drink up to 70 litres in 24 hours. Clean, fresh water must therefore be available at all times; it does not take long for a horse to become dehydrated.

Most horses prefer water that is cold but not icy. In winter, it is vital to break the ice on troughs and containers several times a day. Water in stables absorbs ammonia, so if you use containers rather than automatic waterers it is important to change the supply daily. The drawback with automatic waterers is that they do not allow you to gauge how much the horse is drinking.

Water tastes different in different areas and some horses are reluctant to drink if you stay away from home. Taking a container and mixing in 'home' water with the new supply often helps. If your horse is reluctant to drink at a competition and you are worried that he may become dehydrated, try tips from top endurance riders: some horses will drink water mixed with peppermint cordial or fresh water in which sugar beet pellets have been soaked.

BUCKET TIPS

Research at the Animal Health Trust, Newmarket, has shown that some horses are more ready to drink if their buckets are placed in wall clips rather than on the ground.

STORAGE

There are so many feed companies that choosing the brand to buy can be bewildering. Factors to take into account include availability, how far you have to travel, ease of getting expert advice if you have a problem or query and storage conditions at your supplier's premises.

PROTECT YOUR INVESTMENT

All feed must be kept clean, dry and safe from vermin. At home, this means storing it in containers that rats and mice cannot chew. Galvanised feed bins are ideal if you have several animals but are expensive and perhaps larger than the one-horse owner needs. Metal dustbins are a good alternative – plastic ones are no match for determined rats!

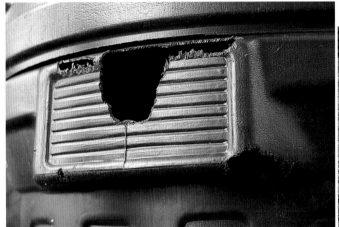

When you finish one batch of feed, make sure the container is clean and dry before tipping in another batch. Old food left in the bottom will go stale and taint the rest. Horses can be fastidious, so wash out mangers after each feed.

Hay should be stored on pallets, preferably in a weatherproof building. Tarpaulins do not really provide enough protection and may make the top layer 'sweat'. If you have plenty of room, using a bottom layer of straw bales also protects your hay.

SUPPLEMENTS AND ADDITIVES

DOES YOUR HORSE NEED THEM?

Supplements and additives should only be given for specific reasons, not because you have been dazzled by clever marketing, or hope that a magic potion will make your horse jump higher! A balanced diet should already provide the correct level of vitamins and minerals. However, if your horse needs less than the feed manufacturer's recommended daily minimum amount – usually about 3 kg, which is a lot for some animals – he may need a broad spectrum supplement at half the recommended dose. Horses fed on straight cereals will need an oat balancer or other form of supplementation to compensate for deficiencies.

Salt is essential for all horses and ponies. Free access to a salt lick is ideal but if the horse will not use it, add 50 g salt per day to his feed. Opinions differ on whether hard working horses need electrolytes – to replace salts lost in sweat – as well as ordinary salt. If electrolytes are offered in water, plain water must also be available.

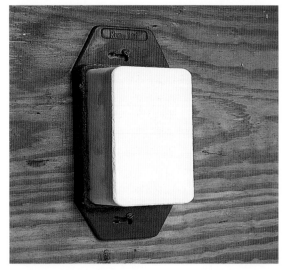

Specialist additives are available for hard-working competition horses but should only be used with expert advice. They are not drugs but, as with all feed substances, competitors must check that they do not contain substances banned by the various disciplines' ruling bodies. Reputable manufacturers will give free advice and information.

Yeasts and probiotics aim to improve digestion and help the horse get the maximum value from his food. They may help with weight gain and are available as separate products or incorporated in feed. Live yoghurt is a probiotic.

HERBS FOR HORSES

When horses have the chance to roam freely and choose their own grazing, they will seek out different herbs; a horse grazed in hand on an established grass verge will pick and choose. The popularity of herbal supplements has boomed over the past few years, though, as with any supplement or additive, they should never be used as a substitute for veterinary advice.

You can plant herbs such as camomile, said to help promote calmness, in your paddock, but you cannot make the horse eat them! Dried herbs are added to some feeds to add palatability and there are specialist mixes designed to help with temperament, respiratory and mobility problems. Many owners report good effects, especially with mixes to promote calmness. Feeding garlic seems to make some horses less attractive to flies.

The biggest criticism about herbal products is that there is no scientific research behind them. However, some of the leading companies are now working with veterinary surgeons and setting up clinical trials. Advocates point out that herbal medicine has been practised successfully for hundreds of years.

WARNING!

Just because something is 'natural', it does not mean it cannot be dangerous. Never exceed recommended doses of any supplement, whether herbal or not, and if you want to use more than one at a time, check that the combination will not cause problems.

ACKNOWLEDGEMENTS

Thanks to Liz Bulbrook, Bailey's Horse Feeds; Ruth Bishop, Spillers Horse
Feeds; Teresa Hollands, Dodson and Horrell and Pippa Stiles, Dengie,
for help in preparing this book.
Centre photograph on page 5 courtesy of the ILPH.

British Library Cataloguing-in-Publication Data.
A catalogue record for this book is available from the
British Library

ISBN 0.85131.707.3

Published in Great Britain in 1997 by
J. A. Allen & Company Limited,
1 Lower Grosvenor Place, Buckingham Palace Road,
London, SW1W OEL

Design and Typesetting by Paul Saunders
Series editor Jane Lake
Printed in Hong Kong by Dah Hua Printing Press Co. Ltd.